Light and

by Susan King

Consultant: Dr. Paul Ohmann, Assistant Professor of Physics,
University of St. Thomas

The sun gives us light.

Lamps give us light.

Candles give us light.

We get light in many ways.

Light can go through glass.

Light can even go through colored glass.

Light cannot go through a soccer ball.

Can you see the ball's shadow?

Light cannot go through me!

Can you see my shadow?

Sometimes a shadow is big.

Sometimes a shadow is small.

My shadow jumps with me.

My shadow runs with me.

What kind of shadow do you see here?